Note: Reproduce this map for your students. See page 2 for ways to use the map.

Continents and Oceans of the World

Continents are large areas of land.
Oceans are large areas of salty water.
Color the water on your map blue.

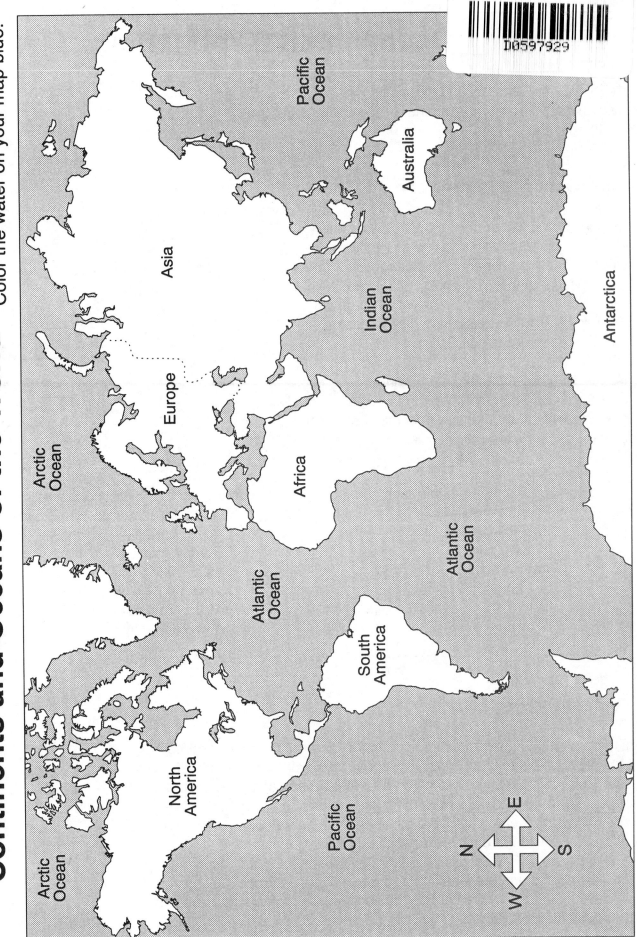

Note: Give each child a copy of the map on page 1.

Oceans and Continents

Have a variety of maps and globes as well as simple atlases available for children to examine as you study about the continents and oceans.

Introduction to Oceans and Continents:

Ask children to tell you what they already know about oceans/continents. Be prepared to explain what an ocean/continent is to children who have had limited experience. (Although it is not included in this unit, some maps list the area where the Atlantic, Pacific and Indian Oceans come together as the Antarctic Ocean.)

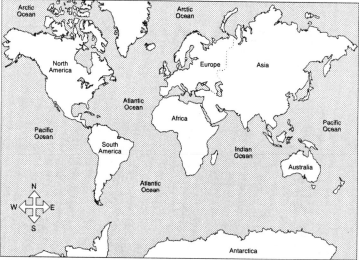

Name each ocean/continent showing its position on a wall map or on the poster in this unit. Have your students mark the area in some way on their own maps. For example...

Underline the name of the each ocean:	Circle the name of each continent:
Atlantic - red	North America - orange
Pacific - green	South America - yellow
Indian - purple	Europe - pink
Arctic - black	Asia - blue
	Australia - brown
	Africa - lavender
	Antarctica - white

Folders and Portfolios:

The map on page one may be glued to the cover of a folder or portfolio. Use this as a place for children to keep all their work for this unit. If you use it this way, have children color the water areas before gluing it to a cover. Then have them color each continent as it is being studied. You may want them to color the continents the same colors as they are colored on the world poster you received with this unit.

Another way to use the map is to reproduce it seven times, having students color in only the continent being discussed each time, leaving the rest of the map white.

This is **North America.**

It is a large continent.
North America has tall mountains.
It has flat plains.
North America has hot deserts.
It has cold glaciers too.

Many people live on this continent.
Do you live in North America?

_____ yes _____ no

Look at the big map.
Find North America.
What oceans touch North America?

_____

_____

_____

Color the animals.
Circle the one that lives in North America.

This is **South America.**

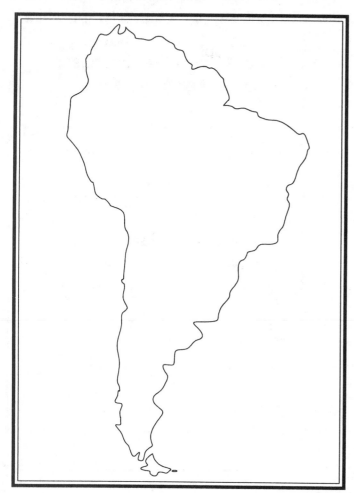

It is a large continent.
There are long mountain ranges.
There are long rivers.
South America has the largest rainforest in the world.
It has flat, grassy areas too.

Do you live in South America?

_____ yes _____ no

Look at the big map.
Find South America.
What oceans touch South America?

_____

_____

Color the animals.
Circle the one that lives in South America.

This is **Asia.**

It is the largest continent.
Asia has many countries.
Some of these countries are islands.
More people live in Asia than on any other continent.
These people speak different languages.
These people live in different ways.

Do you live in Asia?

_____ yes _____ no

Look at the big map.
Find Asia.
What oceans touch Asia?

- - - - - - - - - - - - - - - - - -

- - - - - - - - - - - - - - - - - -

- - - - - - - - - - - - - - - - - -

Color the animals.
Circle the one that lives in Asia.

This is **Europe.**

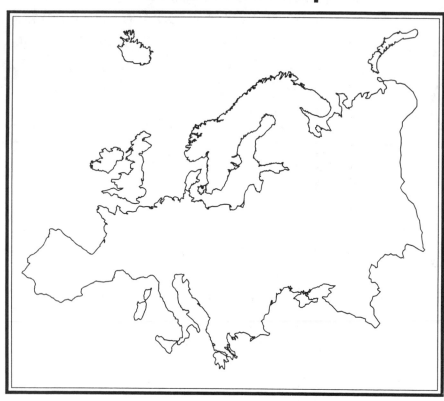

Europe is a small continent.
But it has many countries.
Parts of Europe are very cold.
Parts of Europe are sunny and warm.
Europe has many people living on farms and in big cities.

Do you live in Europe?

_____ yes _____ no

Look at the big map.
Find Europe.
What oceans touch Europe?

Color the animals.
Circle the one that lives in Europe.

This is **Africa.**

It is a large continent.
Africa has the largest desert in the world.
It has the longest river in the world.
There are many different kinds of animals in Africa.
Africa has more than fifty different countries.
The people speak different languages.
They live in different ways.

Do you live in Africa?

_____ yes _____ no

Look at the big map.
Find Africa.
What oceans touch Africa?

Color the animals.
Circle the one that lives in Africa.

This is **Australia.**

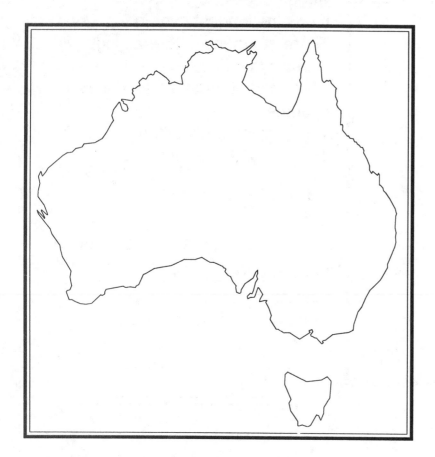

It has water all around it.
Australia has many deserts.
It has rainforests, too.
There are many unusual animals
in Australia.

Do you live in Australia?

_____ yes _____ no

Look at the big map.
Find Australia.
What oceans touch Australia?

Color the animals.
Circle the one that lives in Australia.

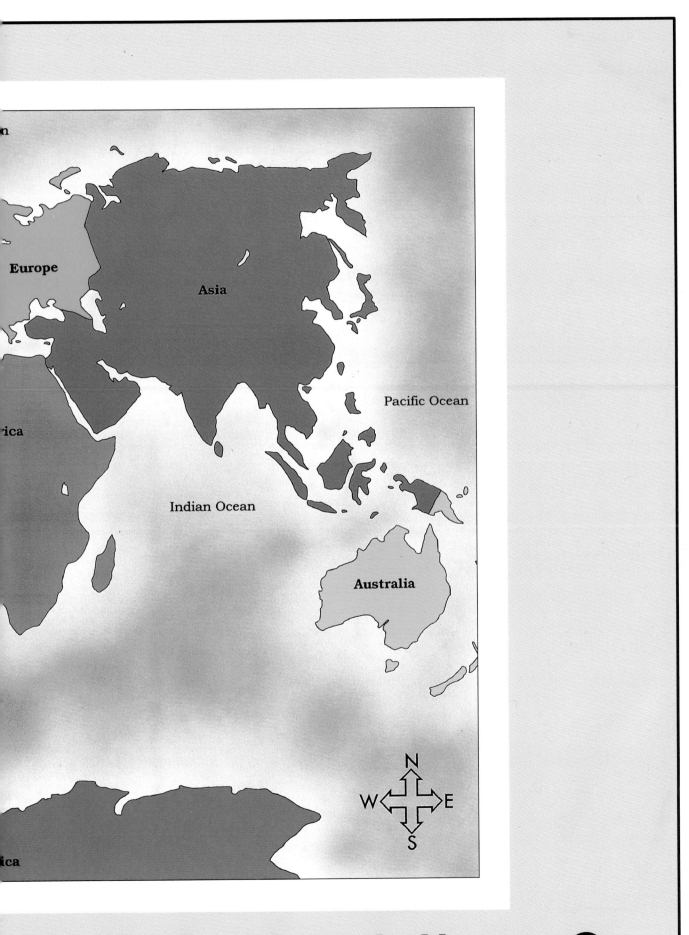

Europe

Asia

Pacific Ocean

ica

Indian Ocean

Australia

N

W ← → E

S

orld do I live?

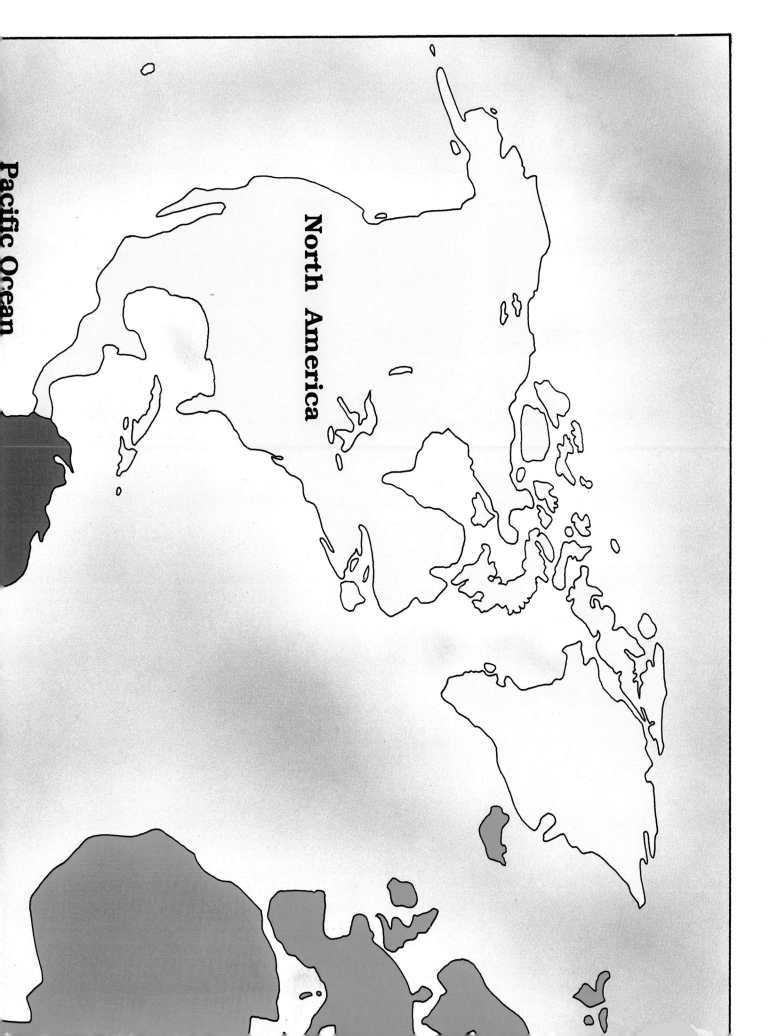

world do I live?

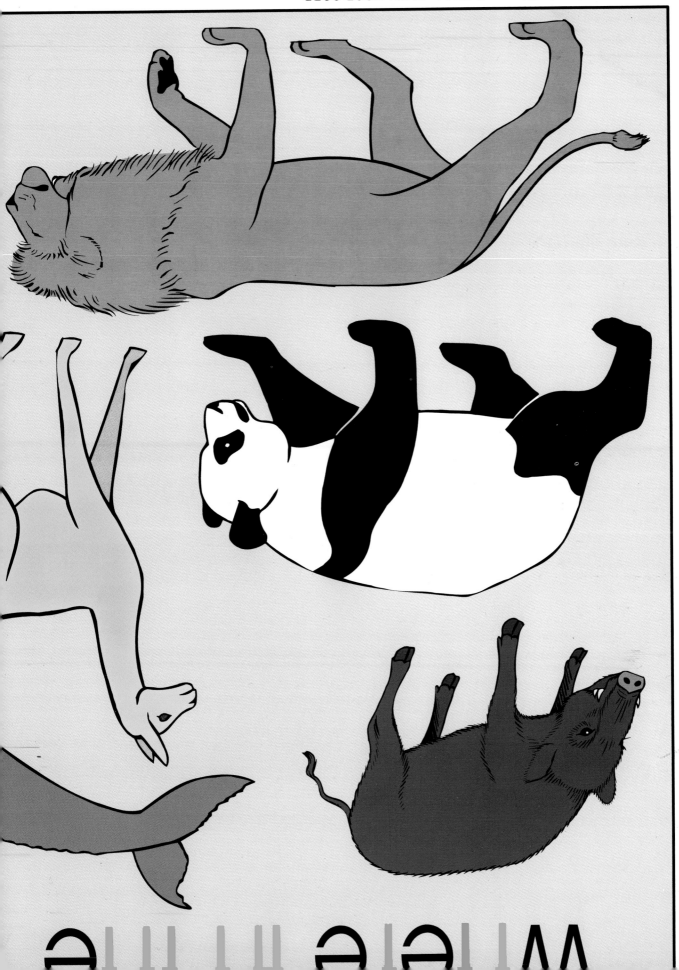

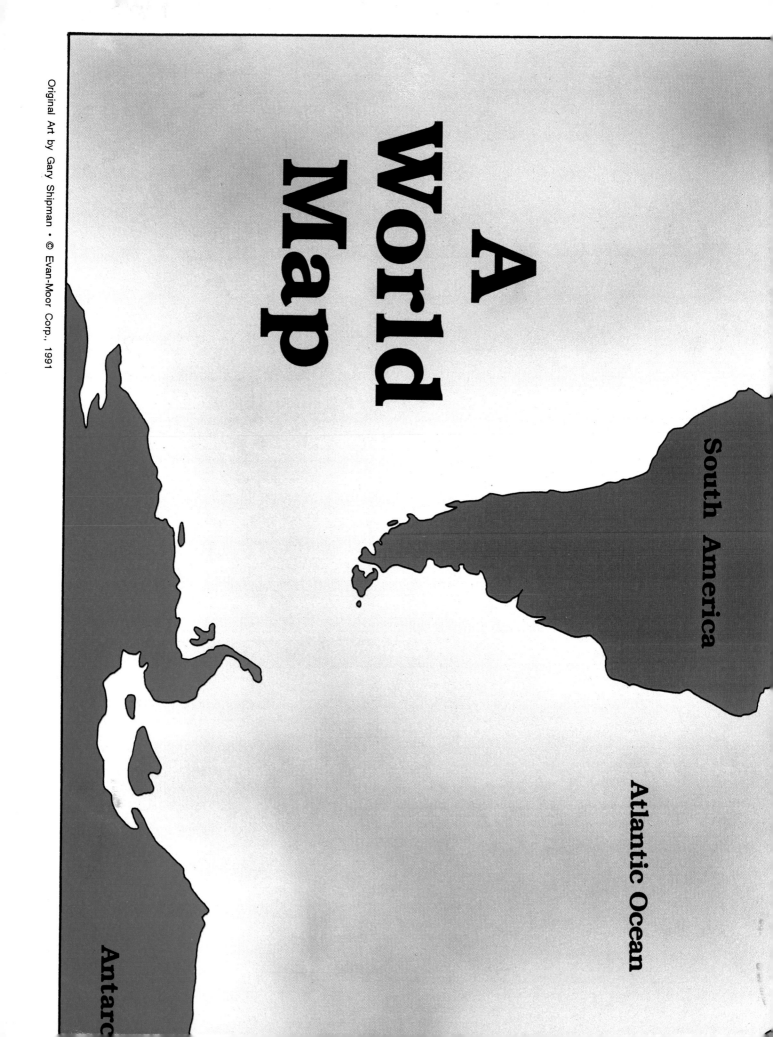

A World Map

South America

Atlantic Ocean

Antarc

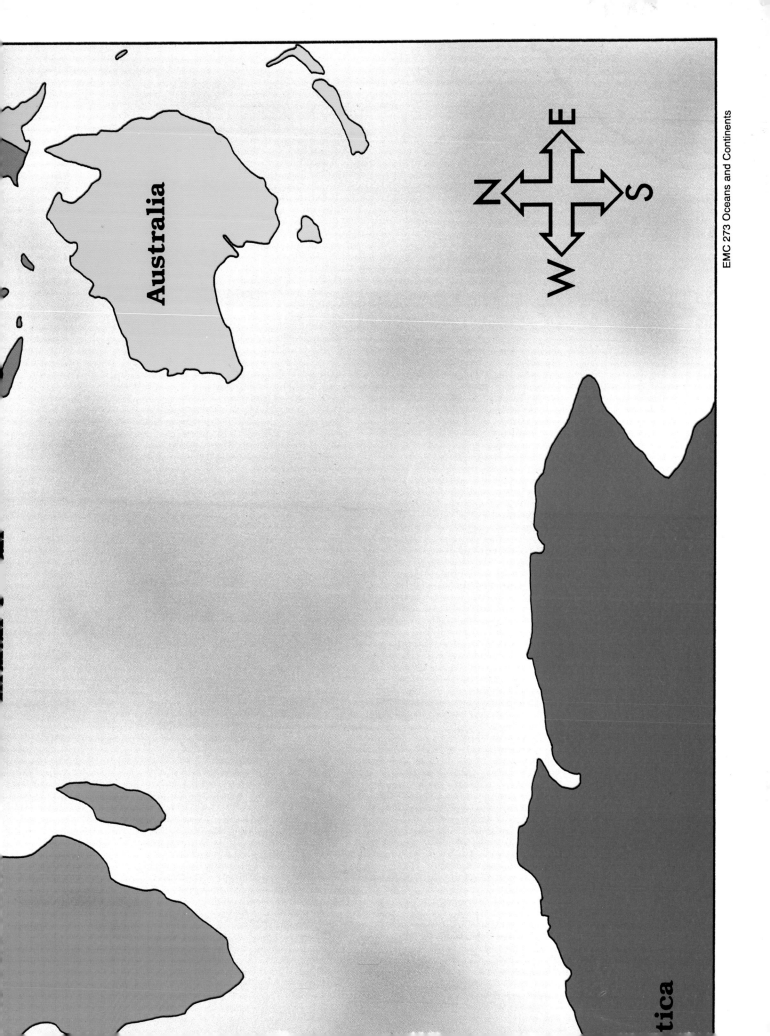

Australia

tica

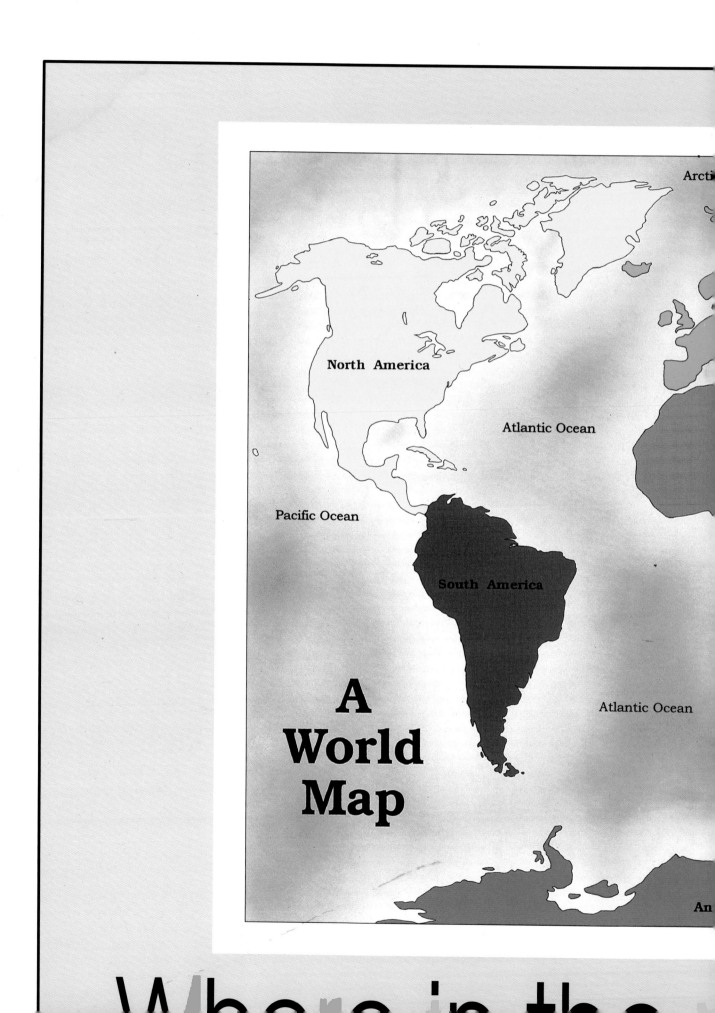

Arcti

North America

Atlantic Ocean

Pacific Ocean

A World Map

South America

Atlantic Ocean

An

Where in the

Antarctica is very cold.

The land is buried under ice.
Only a few small plants grow in
Antarctica.
Only a few kinds of animals live there.

Do you live in Antarctica?

_____ yes _____ no

Look at the big map.
Find Antarctica.
What oceans touch Antarctica?

Color the animals.
Circle the one that lives in Antarctica..

Name the Oceans

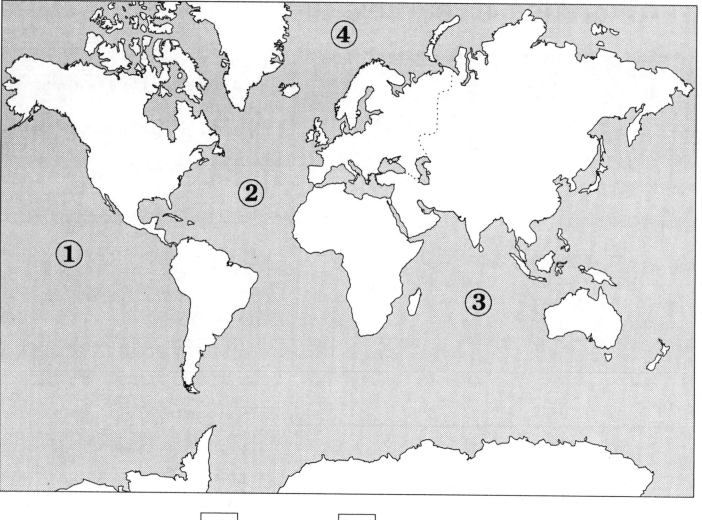

1.

2.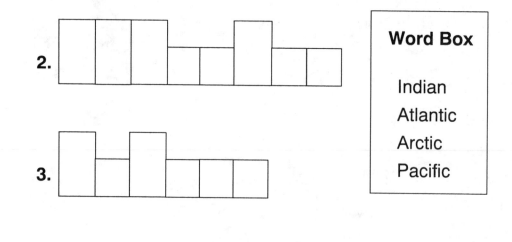

Indian
Atlantic
Arctic
Pacific

3.

4.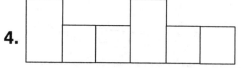

Continents and Oceans

Name the Continents

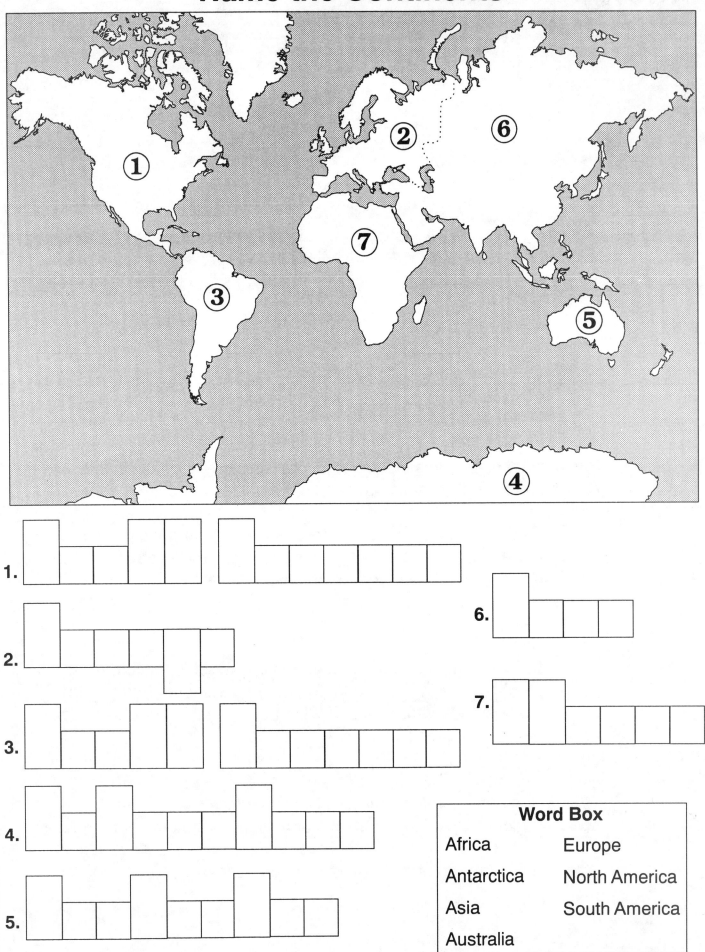

1. ☐☐☐☐☐ ☐☐☐☐☐☐

2. ☐☐☐☐☐☐

3. ☐☐☐☐☐ ☐☐☐☐☐☐

4. ☐☐☐☐☐☐☐☐☐

5. ☐☐☐☐☐☐☐

6. ☐☐☐☐

7. ☐☐☐☐☐☐

Word Box

Africa	Europe
Antarctica	North America
Asia	South America
Australia	

11

Continents and Oceans

North, South, East, West

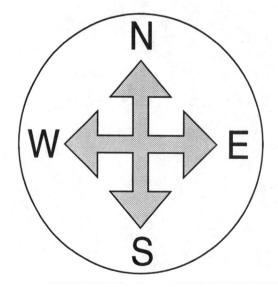

Show your children this picture of a compass rose. Explain its use. Ask them to find the compass rose on the poster, then have them...

- Find the letter N on your compass rose. What does it stand for? Is it at the top, the bottom, on the right or on the left of your compass? (Repeat for south, east and west.)

- Put your finger at the top of your map. Which direction is this? (north) Write an N at the top of the map with your red crayon. (Repeat for south, east and west.)

- Find Asia. Which continent is west of Asia? Find North America. Which continent is south of North America? Find North America. Is Europe east or west of North America? (Repeat with other continents and oceans.)

 Continents and Oceans

Ways to Use Your Continent and Animal Cards

Players: 2

<u>Preparation:</u> Reproduce two copies of the cards on pages 13, 14, and 15 on tag or cardstock. Laminate the pieces for long wear. Mix the cards up. Place them face down on a table. You may want to limit the number of cards for younger players.

<u>Rules:</u> The first player turns over two cards. If the cards match, the player keeps the cards. If the cards don't match they are turned back over. Then the second player takes a turn. The players continue taking turns until all of the cards have been picked up.

Match Up

<u>Preparation:</u> Reproduce one copy of the cards on pages 13, 14, and 15 on tag or cardstock. Laminate the pieces for long wear. Make a set of word cards the same size as your continent and animal cards. Write the names of the continents on the cards. Place the name cards, continent cards and animal cards in an envelope.

<u>Use:</u> Place the envelope of cards in a center. Students then use the cards to practice recognizing and naming the places and animals, and putting the cards together in sets.

You can use these same cards as a quick test by reproducing them on paper. Give each child you are testing a set of the cards. The child cuts them out and places them together in sets.

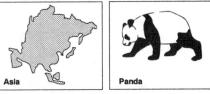

Bean Bag Toss

<u>Preparation:</u> You will need a beanbag and a playing board. Make the playing board by enlarging the continent and animal cards. Glue them to a sheet of butcher paper. Lay the playing board on the ground and draw a line with chalk about six feet away.

Players: 2-4

<u>Rules:</u> The player throws the beanbag to the playing board. He/she must name each continent or animal the beanbag is touching. One point is given for each correct answer. A variation is to use two beanbags. The child tosses the beanbag at a continent. He/she then tries to toss the other beanbag onto an animal found on that continent.

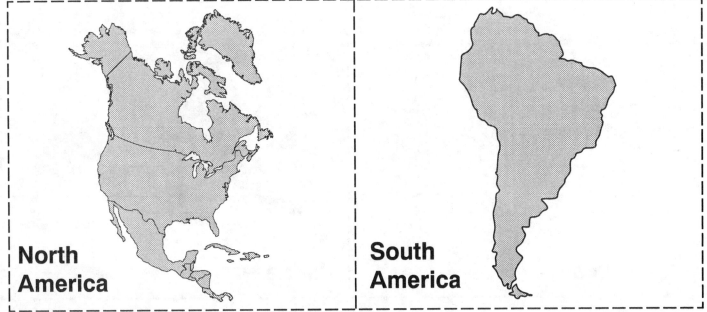

North America

South America

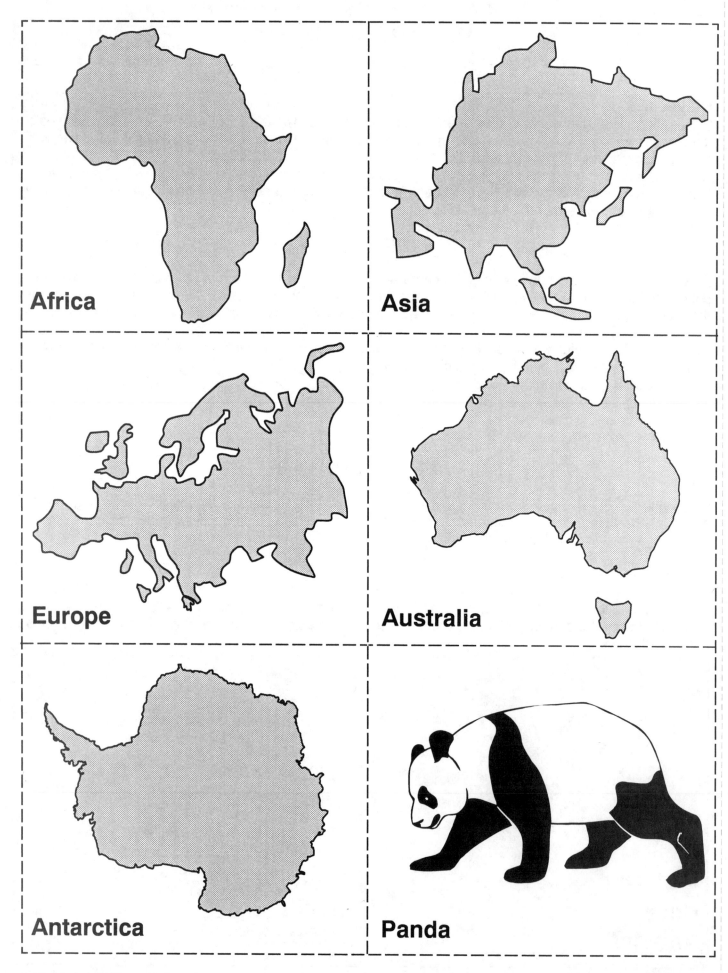

Africa

Asia

Europe

Australia

Antarctica

Panda

Continents and Oceans

Lion

Kangaroo

Buffalo

Llama

Penguin

Wild Boar

Note: Your students will need a copy of this, a sheet of construction paper, scissors and paste.

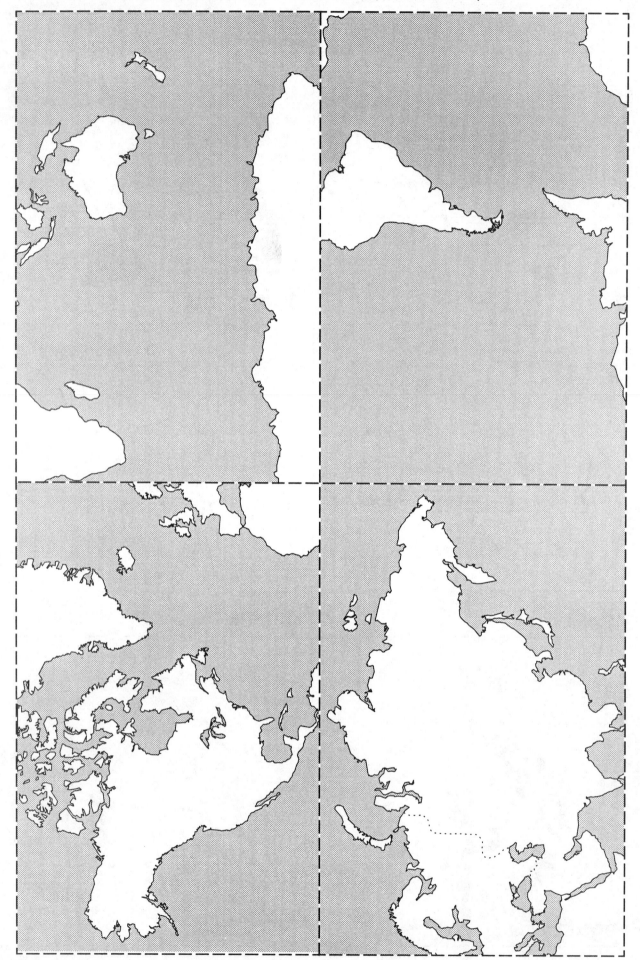

Continents and Oceans